PLAY & LEARN MATH
Number Sense
Learning Games and Activities to Help Build Foundational Math Skills

by Susan Andrews Kunze

SCHOLASTIC
Teacher
RESOURCES

Editor: Maria L. Chang
Cover design by Tannaz Fassihi
Cover art by Constanza Basaluzzo
Interior design by Grafica Inc.
Interior art by Mike Moran

Scholastic Inc., 557 Broadway, New York, NY 10012
ISBN: 978-1-338-64128-8
Copyright © 2021 by Susan Andrews Kunze
All rights reserved.
Printed in the U.S.A.
First printing, January 2021.
1 2 3 4 5 6 7 8 9 10   40   26 25 24 23 22 21

Contents

Introduction

"I get it, Mrs. Kunze!" Jared burst out during our small-group math session. A struggling student, Jared had been playing our game, making groups of tens and ones as we rolled a die and added more counters. At every round, he had difficulty figuring out the "trade" of ten ones into one group of ten. But as he worked building numbers, the pattern eventually became clear. Nothing is more exciting to a teacher than that "ah-ha!" moment when a student makes a connection that leads to understanding.

Play and Learn Math: Number Sense contains a selection of favorite activities designed to help children in the early grades gain understanding about many basic number concepts, guiding them to their own "ah-ha!" moments as they play and learn. Developing number sense is vitally important for young learners. Number sense is the foundation for all higher-level mathematics (Feikes & Schwingendorf, 2008). It is the ability to think flexibly, conceptually, and fluently about numbers. Children who develop strong number sense have an organized conceptual mathematics framework that allows them to understand number relationships and solve problems. They can reason flexibly with numbers, compose and decompose numbers in a variety of ways, make reasonable estimates, and identify unreasonable answers.

The activities in this book have been divided into sections according to the number-sense skill they most directly address. These include counting and ordering numbers, subitizing (quickly identifying the number of items in a small set without counting), patterns, early number operations, and place value. These activities are based upon basic concepts that number sense requires, including:

- understanding quantity
- recognizing that symbols represent quantities
- subitizing
- reasonably estimating larger amounts
- understanding whole/part relationships to compose and decompose numbers
- ordering numbers in a sequence
- using proportional thinking to make number comparisons, such as more than and less than
- identifying and using patterns in number relationships
- understanding the value of each digit in a number

The activities presented in *Play and Learn Math: Number Sense* guide children into deep mathematical understanding in an engaging, hands-on format. Several activities require standard math manipulatives, such as linking cubes (e.g., Unifix cubes), dominoes, playing cards, number and dot (six-sided) dice, and ten-sided dice. Other items, such as dried beans, toothpicks, and buttons, that are easily found at home provide variety and interest in early number-learning tasks. Gather these few supplies, and you'll be ready to start guiding children in powerful mathematics learning. Children will love these activities, and so will you. Let's get playing and learning!

Mathematics Standards Correlations*

Standard	Active Counting	How Many Buttons?	Number Bead Counters	Counting Up	Buzz	Writing Numbers	Building Domino Dot Patterns	How Many Dots?	Build the Number	Dot-Card Concentration	Gumball Grab	Dot-Pattern Plates	Domino Match	Toss and See	Bounce It!	Subitize Ten Bingo	Pattern Trains	Follow the Color	Numbering Squares	Roll and Graph	Toothpick Patterns	Growing Patterns With Linking Cubes	Grab Some Beans	Bean Toss	Subtraction Trains	Many Ways to Say It	Say It Again	Domino Flip	Cards Make 10	Toss and Count	Shut the Box	Skip-Counting Trains	Breaking Trains	Count and Group Tens	Count and Compare	Tens and Ones Toss Up	Make That Number!
KINDERGARTEN																																					
CC.A.1 Count to 100 by ones and by tens.	✓			✓																																	
CC.A.2 Count forward beginning from a given number within the known sequence.	✓			✓																																	
CC.A.3 Write numbers from 0 to 20.						✓																															
CC.B.4 Understand the relationship between numbers and quantities; connect counting to cardinality.			✓				✓	✓	✓	✓			✓	✓	✓	✓				✓																	
CC.B.5 Count to answer "how many?" questions about as many as 20 things in a line or group.	✓	✓	✓					✓			✓	✓		✓	✓	✓																					
CC.C.6 Identify whether the number of objects in one group is greater than, less than, or equal to the number of objects in another group.			✓	✓						✓	✓	✓	✓																								
CC.C.7 Compare two numbers between 1 and 10 presented as written numerals.															✓	✓																					
OA.A.1 Represent addition and subtraction with objects.																							✓	✓	✓		✓			✓							
OA.A.4 For any number from 1 to 9, find the number that makes 10 when added to the given number.																													✓								
OA.A.5 Fluently add and subtract within 5.																							✓	✓	✓		✓	✓		✓							
MD.B.3 Classify objects into given categories; count the numbers of objects in each category and sort the categories by count.																	✓	✓	✓	✓	✓	✓															
G.A.2 Correctly name shapes regardless of their orientations or size.																						✓															
G.B.5 Model shapes in the world by building shapes from components and drawing shapes.																						✓															
GRADE 1																																					
OA.A.1 Use addition and subtraction within 20 to solve problems involving situations of adding to, taking from, putting together, taking apart, and comparing.																							✓	✓	✓	✓	✓	✓	✓	✓	✓						
OA.A.3 Apply properties of operations as strategies to add and subtract.																															✓						
OA.C.6 Add and subtract within 20, demonstrating fluency for addition and subtraction within 10.																							✓	✓	✓	✓	✓	✓	✓	✓	✓						
NBT.B.2 Understand that the two digits of a two-digit number represent amounts of tens and ones.																																		✓	✓	✓	✓
NBT.B.2.A 10 can be thought of as a bundle of ten ones—called a "ten."																																		✓	✓	✓	✓
NBT.B.3 Compare two two-digit numbers based on meanings of the tens and ones digits.																																			✓	✓	✓
GRADE 2																																					
OA.B.2 Fluently add and subtract within 20 using mental strategies.																							✓	✓	✓	✓	✓	✓	✓	✓	✓						
OA.C.3 Determine whether a group of objects (up to 20) has an odd or even number of members by pairing objects or counting them by 2s.																																✓	✓				

Active Counting

There's more to counting than simply reciting number words in sequence. Being able to say the numbers 1, 2, 3, 4, 5 does not mean a child has attached meaning to each number word. These are two separate, but related, skills. However, committing the counting sequence to memory allows children to use that knowledge to build number understanding.

The following activities—which can be done individually, in pairs, or in small groups—use movement to reinforce the meaning of numbers while naming the number words.

HERE'S HOW

When doing the activities below, start by having children count up to where they have difficulty. As they gain confidence with the counting sequence, add one more number. **Note:** These activities can also be adapted for remote learners at home.

- **Stand/Sit:** While holding hands, have partners count and alternate standing and sitting in a chair (or bending forward). For example: *one* (stand), *two* (sit), *three* (stand), *four* (sit).

- **Body tap:** As they count in sequence, have children tap parts of their own bodies. For example, they can tap their heads, shoulders, knees, then toes in sequence as they count, then repeat the tap pattern as they continue counting.

- **March:** While saying the number sequence, have children march in one direction. Then have them turn around and march back while repeating the sequence. You can also have children hop as an alternative.

- **Bounce a ball:** Have children bounce a ball while counting, using one bounce for each number.

- **Jump rope:** Provide a child with a jump rope and have him count with each jump.

- **Play an instrument:** Invite children to play a simple instrument, such as a drum, cymbals, rhythm sticks, or xylophone, or clap hands as they count the number sequence.

- **Count and turn:** For this whole-class activity, have children form a single line. Choose the number sequence for them to count. As they count, have children stamp their feet. When they reach the last number in the sequence, have them emphasize the number by raising their arms in the air and shouting it. Then have them turn around to face the opposite direction and start the counting sequence again.

MATERIALS

- balls
- jump ropes
- musical instruments (drum, cymbals, rhythm sticks, or xylophone)

VARIATION

You can also use these activities to practice counting backward, counting from a number other than 1, or skip counting.

How Many Buttons?

Use this activity to assess if a child can use one-to-one correspondence when counting. Oftentimes young children can recite numbers, but not clearly understand their meaning. Children need to understand that each number name represents a specific amount before they can learn how to represent that amount in different ways.

Buttons are my first choice for counting because as a child I loved to count the buttons in my grandmother's button box. They are a good tool for this activity, but any similar small objects will work.

HERE'S HOW

1. Spread a few buttons on a flat surface. When assessing the number understanding of a very young learner, start with just three buttons.

2. Ask the child to count the buttons, encouraging him to touch each button as he counts aloud.

3. If the child can do this easily, make a new group with five buttons. Again, ask the child to touch each button as he counts.

4. Continue making new groups, increasing the amount by one or two buttons, until the child has difficulty with counting or naming the number. Use this number as the starting point when doing counting activities with this child.

MATERIALS

• buttons (or other small objects)

Number Bead Counters

These adult-created learning tools are made of pony beads tied together in a way that allows them to be grouped. They are especially useful for unit counting and for demonstrating early addition and subtraction relationships. Make these for young learners to keep on hand for solving problems in early operations.

NOTE: Number Bead Counters may take a bit of time and effort to make, but their usefulness makes them a worthwhile project. They can be assembled at home, so parents who are unable to volunteer in the classroom may be interested in making these at home for your class.

HERE'S HOW

1. Tie a 15-inch waxed linen cord mid-length to a lanyard clip or split ring.

2. Thread the two ends of the waxed cord through the center of the first bead, coming from opposite sides, so that they cross inside the bead. Then pull the ends through completely.

3. Place a second bead of the same color next to the first bead and repeat step 2.

4. Continue adding beads in this manner, putting five of one color in a row, then five of the second color. (The two bead colors represent groups of five.)

5. When all ten beads are attached, make a square knot on the cord, tightly tying the ends as close to the tip of the cord as possible. The knot should be about an inch from the last bead.

6. Make sure the beads can move easily and independently from one another. They should be easy to move in groups and stay in that position.

Show children how to use the Number Bead Counters to count. As they begin early addition, they can manipulate the beads to represent a number sentence, such as 3 + 7 = 10 by moving three beads into a group and leaving seven beads in the other group. Demonstrate the commutative property of addition simply by flipping over the counter.

MATERIALS

(for each number bead counter)

- **15-inch waxed linen cord**
- **1-inch lanyard clip or a tiny split ring**
- **10 pony beads (5 each of 2 different colors)**

Counting Up

As young learners begin to put numbers together in addition, counting objects by ones starting with the first object becomes a very inefficient strategy. Guide children to identify a small number of objects in a group, then to count up from there to find the total number of objects. Extend this teacher-directed activity to partner play as children become more familiar with the concept of counting up.

HERE'S HOW

Use objects that are small enough to hide under your hand or a cup.

1. To start, take a few objects and lay them in a row.

2. Ask a child to count the objects (for example, five objects), then cover two of the objects with your hand.

3. Demonstrate how to count from that point in the number sequence: "*Two ... three, four, five.*"

4. Next, cover three of the objects and demonstrate how to count up from that point in the sequence: "*Three ... four, five.*"

5. Continue with various numbers to emphasize how to count up.

MATERIALS

- small objects, such as buttons, beans, or paper clips
- cup or index card (large enough to cover some of the objects)
- number cube or dot die (optional)

VARIATION

As children gain proficiency with counting up, you can adapt this activity into a game for partners. When working with numbers greater than six, use game elements, such as a number cube or dot die, to increase child interest and engagement. For example, children can roll the number cube to determine how many objects to cover.

Play & Learn Math: Number Sense © Susan Andrews Kunze, Scholastic Inc.

Buzz

"Buzz" is a low-prep, whole-class game that gives children practice in skip counting. They literally have to think about multiples on their feet! Children can play this game using any multiple, so they can play it often without it feeling repetitive. "Buzz" will become a class favorite!

HERE'S HOW

Invite children to stand and form a large circle. Decide and tell children what the skip-counting number (or multiple) will be.

1. To play, children take turns counting around the circle by ones.

2. When a child comes to the multiple, he or she says, "*Buzz*," instead of the number. For example, if skip counting by 5s, children start counting, "*1, 2, 3, 4, Buzz, 6, 7, 8, 9, Buzz*," and so on. If a child forgets to say, "*Buzz*," he or she sits down.

3. Counting continues until only one child is left standing.

PLAYERS   Whole class

MATERIALS

• none

VARIATIONS

• Instead of always starting at 1, choose a different number to start counting.

• As children become more proficient in counting, play may end up lasting longer than class time allows. To avoid this, choose an ending number so children count until they reach that number.

• For a greater challenge, have children start from a larger number and count backward to one.

• Another way to provide challenge is to have children play using two multiples, such as 2 and 5, while skip counting.

Writing Numbers

Writing numbers is an essential early math skill. As children learn to form numbers, provide them with a variety of practice opportunities. Keep examples of perfect number formations so children can check their writing as they work.

Try any of these great ways for children to practice writing numbers. **Note:** These activities can also be done at home by remote learners.

- **Whiteboards/chalkboards:** Using dry-erase markers or chalk, have children write numbers on their boards. Keep some old, clean socks available for erasing if erasers aren't available.

- **Finger paints:** Provide paper and paints at a center for children to finger-paint their numbers.

- **Salt trays:** Pour salt into rimmed cookie sheets, small cafeteria trays, or rimmed paper plates. Have each child trace numbers with a finger on the salt, so that the tray shows the number shape through the salt.

- **Water painting:** Do this fun outdoor activity on a sunny day. Give each child a container of water and an old paintbrush. Have children use the water to "paint" the numbers on the pavement. The sun will naturally dry them up and clean up the "mess."

MATERIALS

- whiteboards and dry-erase markers, or chalkboards and chalk
- old, clean socks (for erasers)
- paper
- finger paints
- rimmed cookie sheets or paper plates
- salt
- water
- paintbrushes
- lemon juice
- iron (for adult use only)
- geoboards
- large rubber bands
- play dough
- cookie dough

Writing Numbers

(continued)

- **Lemon-juice painting:** This activity takes a bit more time, but it's a fun way to have a record of children's work. Distribute clean, small paintbrushes, lemon juice in small cups, and sheets of paper. Using the lemon juice, have children paint their numbers on the paper. Let the paper air-dry, then press a warm iron on the paper. The numbers will "magically" appear.

- **Geoboards:** Give children large rubber bands to form numbers on geoboards. Model how to place a rubber band over the first nail, then wrap the band around various nails to form the number. To remove safely and avoid flying bands, show children how to place a finger over the first nail they used, then carefully lift the rubber band off the other nails.

- **Play dough:** Invite children to "sculpt" numbers out of play dough.

- **Cookie dough:** Children can form cookie dough into number shapes, as well. Make sure children wash their hands before they touch the dough. Collect cookie-dough numbers on a baking tray to bake later. Children can eat their handiwork the following day.

Building Domino Dot Patterns

Subitizing is the ability to identify how many items there are at a glance, without counting. Building domino and dot-dice patterns is a great way to help children build subitizing skills. When children see how the patterns on dominoes and dice develop, they can begin to identify the set of numbers without having to count each dot. This whole-class lesson teaches how to make a domino or dot-dice pattern on a 3-by-3 grid.

Give each child a copy of the 9-Dot Frame and nine plastic chips. On the board, demonstrate how to build the domino (or dice) dot patterns using the 9-Dot Frame (see below). Invite children to build each number as you demonstrate the sequence. Repeat the entire sequence several times until children understand the pattern.

1. Place one chip in the center space to represent the number 1.

2. Move the center chip to one corner. Add another chip to the opposite corner to make 2.

3. Place one chip in the center to make 3.

4. Move the center chip to another corner. Add another chip to the opposite corner to make 4.

5. Place one chip in the center to make 5.

6. Move the center chip to one side. Add another chip to the opposite side to make 6.

7. Place one chip in the center to make 7.

8. Move the center chip to the top. Add another chip to the bottom to make 8.

9. Place one chip in the center to make 9.

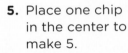
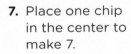
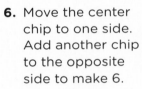
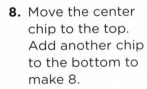

Guide children to notice the pattern of first putting one chip in the center, then moving it out of the center and adding one more.

MATERIALS

(for each child)

- **9-Dot Frame*** (page 15)
- **9 plastic chips**

* For durability, photocopy the 9-Dot Frame onto cardstock and laminate.

VARIATIONS

- To make domino patterns to 18, use an 18-Dot Frame (page 16). Follow the same pattern of putting one marker in the center, then moving it out of the center and adding one more.

- Use the 18-Dot Frame to create a modified domino pattern for the numbers 1–10. Guide children through building the pattern for 1–5 on the top 3-by-3 grid, then add chips one at a time on the bottom array as a "5 plus some more" concept. In the bottom half ("plus some more"), add chips directly to their correct position, rather than moving into position from the center.

9-Dot Frame

Use plastic chips to build domino dot patterns. Follow your teacher.

18-Dot Frame

Use plastic chips to build domino dot patterns. Follow your teacher.

Play & Learn Math: Number Sense © Susan Andrews Kunze, Scholastic Inc.

How Many Dots?

Give young children practice in subitizing groups of dots with this fun activity. As children begin to recognize familiar arrangements of dots, they become more adept at identifying quantities in a set without counting.

HERE'S HOW

Partner up children. Based on their skill level, decide which Dot Cards to provide children. Give each pair two sets of Dot Cards. Give each child a whiteboard, a dry-erase marker, and eraser.

1. Shuffle the Dot Cards and stack them facedown between partners.

2. One child takes a card from the stack and puts it down where both partners can see it.

3. Both children write the number on their own whiteboards. Then they exchange boards to compare their answers.

   • If both agree on their answers, they can erase their boards and continue playing with the other child taking a card from the stack.

   • If they disagree, have them count the dots on the Dot Card to check their answers.

4. Partners continue taking turns picking a card and writing the number as long as time allows.

MATERIALS

• Dot Cards* (page 63)
• whiteboards
• dry-erase markers and erasers

* For durability, photocopy the Dot Cards onto cardstock and laminate before cutting apart. You'll need a set of Dot Cards for each child.

VARIATION

For remote learners, the teacher displays a Dot Card on her screen and children write the number on paper to show the teacher on their screen.

Build the Number

As children replicate the arrangement of dots on various Dot Cards, they build skills in counting and identifying equal amounts. While the materials call for linking cubes, feel free to use any small items you may have available, such as pom-poms, mini erasers, or small toys.

HERE'S HOW

Determine which Dot Cards to give children before they play the game. For younger learners, start with the numbers 1 to 6. Over time, build up to 10 or 12.

1. Divide the class into small groups of three or four children. Give each group the same number of Dot Card sets as the number of children in the group (for example, a group of four children should have four sets of Dot Cards).

2. Assign one player to be the "dealer." The dealer shuffles the Dot Cards and deals one card to each player.

3. Using linking cubes, players copy the arrangement of dots on their cards.

4. The dealer checks each player's result and points out any mistakes for the player to correct.

5. The dealer then deals out new cards to the players, placing the new card on top of the previous card.

6. Players rearrange their linking cubes to match the new card, taking or removing cubes as needed.

7. Play continues until all cards have been used.

PLAYERS 3 or 4

MATERIALS

(for each player)

- Dot Cards* (page 63)
- linking cubes (or other small objects)

* For durability, photocopy the Dot Cards onto cardstock and laminate before cutting apart. You'll need a set of Dot Cards for each child.

VARIATION

For remote learners, have children go on a scavenger hunt to find small objects, such as buttons or beans, in their homes. The teacher can display a Dot Card on her screen for children to copy the arrangement of dots using their objects.

Play & Learn Math: Number Sense © Susan Andrews Kunze, Scholastic Inc.

Dot-Card Concentration

A simple game of Concentration offers children fun practice in matching numbers with their quantities. Based on their skill level, decide how many and which cards to provide players. Early learners can start with cards up to the numbers 5 or 6. Over time, children will successfully identify larger quantities.

HERE'S HOW

Partner up children and give each pair a set of Dot and Number Cards, based on their skill level.

1. Shuffle the Dot and Number Cards together and lay them facedown in a rectangular array.

2. Player 1 turns over any two cards in the array.

   • If the two cards are a matching pair of number and dots, the player takes and keeps them.

   • If the cards do not match, the player turns the cards facedown.

3. Player 2 takes a turn, repeating step 2.

4. Players continue taking turns turning over cards until they have matched all the Dot and Number Cards.

5. Players count their cards. The player with the most matches wins.

PLAYERS 2

MATERIALS

• Dot Cards* (page 63)
• Number Cards* (page 64)

* For durability, photocopy the Dot Cards and Number Cards onto same-color cardstock and laminate before cutting apart.

Play & Learn Math: Number Sense © Susan Andrews Kunze, Scholastic Inc.

Gumball Grab

In this quick and easy game, children pick a Number Card, then build that number of "gumballs" on their game board. The game provides practice in counting numbers and comparing amounts. Use a penny toss to randomize which amount wins and keep children engaged in the activity.

HERE'S HOW

Partner up children and provide each pair with two sets of Number Cards. Give each player a Gumball Game Board.

1. Shuffle all the cards and stack them facedown between the players.

2. To play, each player takes a Number Card from the stack and places it faceup on his or her game board. Players then use plastic chips to build that amount in their gumball machine.

3. Players check each other's game board for accuracy.

4. To determine who wins the round, one player flips a penny. "Heads" means the player with the greater amount wins. "Tails" means the player with the lesser amount wins. The winner keeps one chip in his or her win pile.

5. Play continues until time runs out. The player with the most chips at the end of the game wins.

PLAYERS 2

MATERIALS

- **Number Cards***
  (page 64)
- **Gumball Game Board***
  (page 21)
- **plastic chips**
- **penny**

* For durability, photocopy the Gumball Game Board and the Number Cards onto cardstock. Laminate and cut the cards apart. You'll need a set of Number Cards for each player.

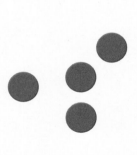
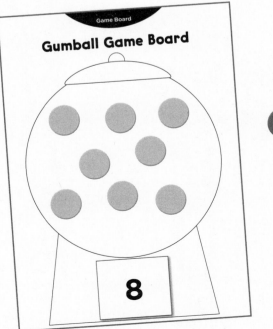

Gumball Game Board

8

Gumball Game Board

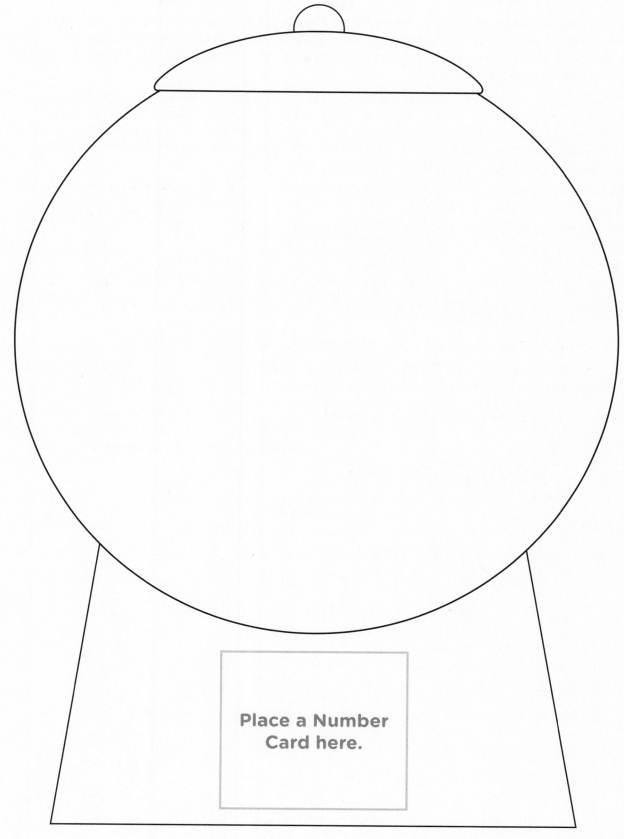

Place a Number
Card here.

Dot-Pattern Plates

Subitizing is the ability to quickly identify the number of items in a small set without counting. Being able to subitize means children aren't stopping to count members of a set one-by-one, because they are able to see a small set as a complete amount or a combination of smaller amounts.

There are two types of subitizing. *Perceptual subitizing* is the ability to recognize a complete amount, such as a group of three or four items. *Conceptual subitizing* is being able to see parts of the set and put them together. For example, children might see a set of seven dots as three dots and four dots or as three dots, two dots, and two dots. They can then quickly add those to find the sum.

There is a strong correlation between subitizing and math achievement. Research shows that subitizing combines with other mental processes to form a foundation for number learning and supports the development of numeration and calculation. Guiding children to develop their subitizing skills helps them become nimbler with numbers.

Many subitizing activities often use dot patterns, commonly found on dice or dominoes. In the classroom, you may also use ten-frame patterns. All these familiar patterns are very helpful in developing children's subitizing skills. Using unusual patterns, however, provides children more conceptual subitizing practice.

HERE'S HOW

Make your own unique dot patterns so children associate more than one pattern with any given number. A quick and simple way to do this is to make Dot-Pattern Plates. Using felt-tip markers or dot stickers on large paper plates, make dot patterns for the numbers 0 to 12. Make sure the dots appear in rows or columns of four or less so children can identify them by sight. (See right for examples.)

1. One way to use Dot-Pattern Plates is to "flash" them one at a time and have children call out the amount they see in the set.

2. Another way to play is to make several sets of 1–12 Dot-Pattern Plates, using different dot patterns for each set. Distribute the plates, one to each child, then call out a number. Invite children with that amount on their plates to dash up to the front of the room and show them to the class. This provides children an opportunity to see a variety of patterns showing the same amount.

With either activity, take a moment to discuss what patterns children noticed as they subitized the set.

MATERIALS

- large paper plates
- felt-tip markers or dot stickers

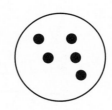

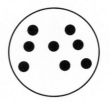

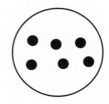

Domino Match

Children match a number with the correct number of dots on a domino. The purpose is to subitize small groupings of numbers, not to add. So players look only for the chosen number on one half of each domino.

HERE'S HOW

Partner up children and give each pair a set of double-six dominoes. Let each player choose which color of plastic chips to use.

1. Players lay out the dominoes faceup between them.

2. Players take turns rolling the number cube to get a number.

3. Both players look for domino halves with that matching number of dots. Each player then covers one domino half with his or her plastic chip.

4. Play continues with players taking turns rolling the number cube until all domino halves have been covered or time has run out.

5. When all domino halves have been covered, players count the plastic chips in their color. The player who used the most chips wins.

PLAYERS 2

MATERIALS

(for each pair of players)

- set of double-6 dominoes
- plastic chips in 2 colors
- number cube

VARIATIONS

- To simplify the game, use fewer dominoes. Be sure the set of dominoes has the numbers 1 to 6 represented.

- To support struggling learners, have two children play together as a team against another team.

- Don't have any number cubes? Give each pair of players a small stack of Number Cards (page 64) from 1 to 6. Shuffle the cards and stack them facedown between players.

- To play with numbers to 10, use double-nine dominoes and a ten-sided die (decahedron). If the die lands on 10, roll again to get a different number.

Toss and See

"Toss and See" provides fun, active practice in subitizing small quantities of items. Gather hula hoops and beanbags from your phys-ed supplies, and watch children have a blast with this game.

HERE'S HOW

For each group of players, lay a hula hoop on the floor. Use masking tape to mark a throw line several feet from the hoop.

1. Players take turns rolling the ten-sided die to determine how many beanbags to use.

2. Standing behind the throw line, a player gently tosses all the beanbags, aiming to get them inside the hoop.

3. Without counting, the other players try to identify the number of beanbags inside and outside the hula hoop.

4. Game continues with players taking turns rolling the die, tossing the beanbags, and calling out the total numbers of beanbags inside and outside the hoop.

PLAYERS   4 to 6

MATERIALS

(for each group of players)

- hula hoop*
- masking tape
- decahedron (10-sided) die
- 10 beanbags

* No hula hoops? Use a long piece of yarn to form a circle on the floor. If you don't have beanbags, try soft whiteboard erasers, large pink rubber erasers, or clean socks rolled into balls.

Play & Learn Math: Number Sense © Susan Andrews Kunze, Scholastic Inc.

Bounce It!

In this fast-paced, two-player board game, children take turns rolling a die to get a number and covering that number on the board. If another player's marker is there, simply "bounce" it off the board!

HERE'S HOW

Decide on which game board and type of die to provide players, based on their skill level. Let each player choose which color of plastic chips to use.

1. Players take turns rolling the die and using a plastic chip to cover a space on the game board with the matching number.

   - If a player rolls a number that already has a chip on it, she can "bounce" off (or remove) that chip and replace it with her own chip.

   - If a player rolls a number that is already covered with his own chip, the player can add a second chip of the same color and stack it on the first one.

   - If a number already has a two-chip stack, that stack cannot be bounced off. The player must either find another space with the same number on which to place her chip or lose a turn.

2. Play continues with players taking turns until all the spaces on the board are covered.

3. Each player counts how many plastic chips he or she has on the board. The player with the most chips on the board wins.

PLAYERS 2

MATERIALS

- **Bounce It! Game Board*** **(pages 26–28)**
- **number cube, dot die, or decahedron (10-sided) die***
- **clear plastic chips in 2 colors**

* There are three different game boards, and each one requires a different type of die. Photocopy the appropriate game board and laminate for durability.

Bounce It!
Numbers 1–6

Use a dot die with this game board.

Play & Learn Math: Number Sense © Susan Andrews Kunze, Scholastic Inc.

Bounce It!
Dots 1–6

Use a number cube with this game board.

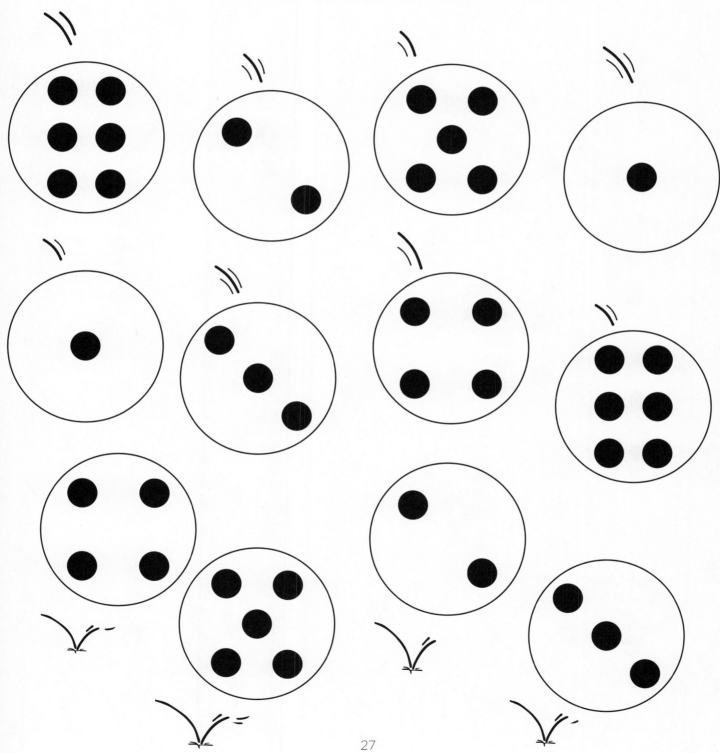

Bounce It!
Dots 1–10

Use a decahedron or ten-sided die with this game board.

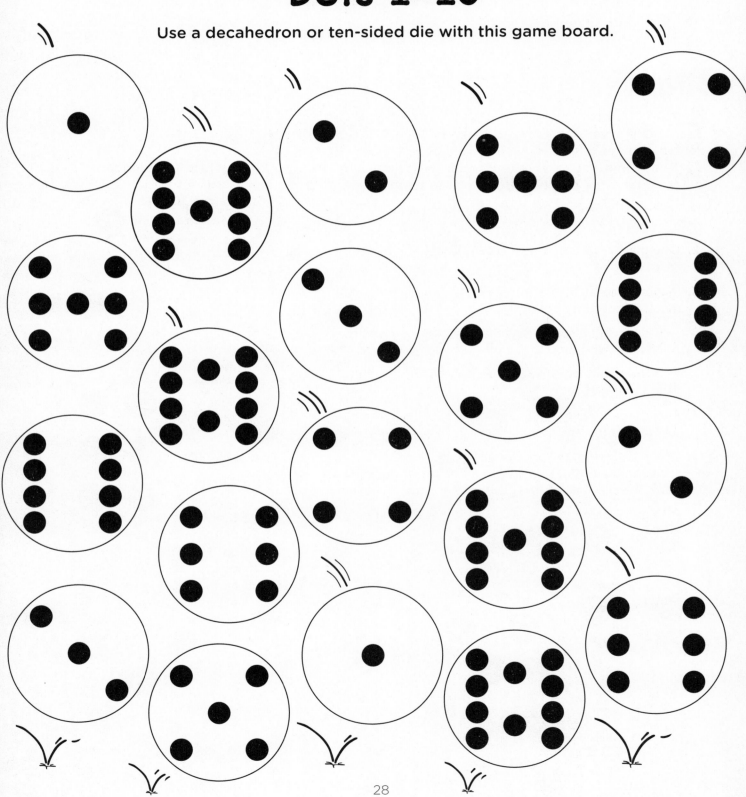

Play & Learn Math: Number Sense © Susan Andrews Kunze. Scholastic Inc.

Subitize Ten Bingo

The bingo cards in this game display numbers in several ways: as dot-dice and domino patterns, on a ten frame, on a *rekenrek* (number rack, counting the beads on the left side of the rack), as a linking-cube train, and as tally marks. Watch children's subitizing skills soar after playing this game a few times.

HERE'S HOW

Photocopy a set of the four Subitize Ten Bingo Cards for each group of players. The cards are labeled A to D, so each player can have a different card. Before children start the game, you may want to review with them the different ways the numbers are displayed on the cards.

1. Give each player a different bingo card and 25 plastic chips.

2. One player rolls the die and calls out the number rolled.

3. Players look at their bingo cards to find a space that matches that number.

   • If they find a space that matches the number, they cover it with a plastic chip.

   • If their card doesn't have a space with the matching number, they lose a turn.

4. Play continues with players taking turns rolling the die and covering the matching spaces on their boards.

5. When a player covers five squares in a row (horizontal, vertical, or diagonal) on his or her bingo card, the player calls out, "Bingo!" and wins the game.

6. Players clear their boards and start a new game.

PLAYERS 2 to 4

MATERIALS

- **Subitize Ten Bingo Cards A to D (pages 30–33)**
- clear plastic chips
- **decahedron (10-sided) die***

* If you don't have a decahedron die, give each group of players a small stack of Number Cards (page 64) from 1 to 10. Shuffle the cards and stack them facedown between players.

VARIATION

💻 For remote learners, give each child all four Bingo Cards and let children choose one card to play. Play as directed but with the teacher rolling the die and calling out the number.

Subitize Ten Bingo Card A

⬤⬤ ⬤ ⬤⬤	ten frame	abacus	| | | |	number rod
abacus	tally marks	number rod	⬤⬤⬤	ten frame
number rod	⬤⬤ ⬤⬤ ⬤⬤	**FREE SPACE**	tally marks	abacus
ten frame	number rod	tally marks	abacus	⬤⬤ ⬤⬤
⬤ ⬤	abacus	ten frame	tally marks	number rod

Subitize Ten Bingo Card B

			FREE SPACE	

Subitize Ten Bingo Card C

Subitize Ten Bingo Card D

Pattern Trains

Make repeating patterns visual for young learners by having them build "trains" using different colors of linking cubes. These patterns grow in a linear manner, repeating the changing colors again and again. Children can start a simple pattern using two or three colors of cubes, then repeat that identical pattern, and eventually connect several of these groupings to make a train.

HERE'S HOW

Children can work individually or with a partner to make simple repeating patterns with linking cubes of two or three colors. Some patterns for children to build include:

- AB
- AAB
- AABB
- ABB
- ABC
- ABBC
- AABCC

1. Assign a pattern for children to build. Provide linking cubes of various colors and allow children to choose the colors to use to build each pattern "train."

2. Encourage children to compare their pattern trains with a partner.

   Initially, children may describe the pattern they built by the color of each cube; for example, "red, yellow, red, yellow, ..." As they become adept with identifying the pattern, encourage them to start describing it using letters. Teach them to designate each new color in the pattern with a new letter, starting with A for the first color used. So a pattern described as "blue, orange, blue, orange, blue, orange" becomes "ABABAB."

3. Try this activity many times, using a different pattern with each opportunity. As children make their pattern trains using different colors, discuss the patterns, using color names and letters to describe each pattern.

MATERIALS

- linking cubes in 2 or 3 different colors

VARIATION

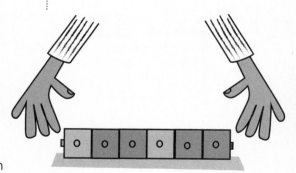 Remote learners can use toy building bricks or blocks instead of linking cubes. They can lay the bricks in a row, not connected.

Follow the Color

As children become more comfortable with creating various linear patterns with colored linking cubes, or pattern trains (see page 34), extend the activity by having them break apart their long train into smaller, equal-size groups. Stacking these smaller groups creates a different kind of pattern. This activity provides a challenge in identifying patterns, as well as a structure for recording pattern information.

HERE'S HOW

Partner up children and give each pair linking cubes in two colors, a number cube, and a large sheet of paper and pencil.

1. Have partners use the linking cubes to build an AB pattern train. The train should be 25 to 50 cubes long.

2. Next, have one child roll the number cube to determine how to group or break apart the linking-cube train. For example, if a child rolls a 4, partners should break apart their train every four cubes.

3. Then, have partners lay the smaller groups in horizontal rows, one below another, on their paper.

4. Tell children to label the grouping number (in our example, 4) on the paper. After choosing which color pattern (A or B) to record, children use arrows to draw the color pattern in the grouping below the number.

MATERIALS

- **linking cubes in 2 colors**
- **number cube or dot die**
- **large sheets of paper and pencils**

VARIATIONS

- To explore larger numbers, have children use a decahedron (ten-sided) die to determine group length.

- Challenge children to use a different pattern when building trains (for example, ABB or AAB patterns). Record the patterns for the number rolled and compare the patterns.

- ⌨ Remote learners can use toy building bricks or blocks instead of linking cubes. They can lay the bricks in a row, not connected.

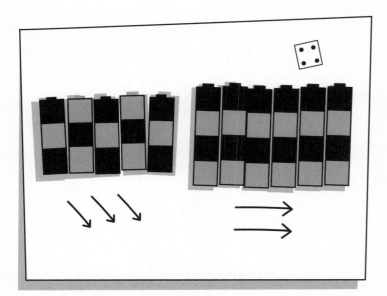

Numbering Squares

In this activity, children practice writing short sequences of numbers on graph-paper squares. As they complete each array, or set of squares, children get an opportunity to look for and make use of structure by observing patterns and extending them to make new understandings. Encourage them to look closely at each array to discern a pattern. Use this activity with a variety of number sequences over time to encourage pattern-seeking and math discussion.

HERE'S HOW

Photocopy the Numbering Squares Arrays or make your own master copy. To make a master copy, outline several arrays (for example, 2-by-2, 3-by-3, 4-by-4, and so on) on a sheet of graph paper. Photocopy the array page for each child.

1. Choose the number sequence you want children to practice writing. If they are practicing counting to 4, for example, the sequence they'll write is 1, 2, 3, 4. Children will write this same sequence of numbers to fill in each array on the page.

2. Starting in the upper left-hand corner of an array, children write the number sequence across the top row. When they come to the end of a row in the array, they should write the next number in the sequence to start the next row. When the number sequence ends, children start the sequence again from the beginning. (See example below.)

3. Have children continue writing the number sequence repeatedly until they completely fill in the array.

4. Then have them pick another array to start writing the number sequence again.

5. After they have filled in all the arrays, have children highlight (or lightly color) all the squares with the number 1 in them.

6. Invite children to talk about the number patterns they have discovered in their arrays.

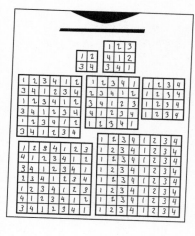

MATERIALS

- **Numbering Squares Arrays* (page 37)**
- **pencils**
- **highlighters or other coloring tool**

* Alternatively, you can use plain graph paper with squares large enough for children to write a number in each square—about $\frac{1}{2}$ inch.

VARIATIONS

- Older children (in second grade and above) often have experience using arrays to represent multiplication problems, so they can make their own array pages on graph paper.

- Use any sequential number pattern to fill the arrays. For more challenge, have children fill in the arrays skip-counting by 2s or 3s. Color in the smallest number in each sequence.

Play & Learn Math: Number Sense © Susan Andrews Kunze, Scholastic Inc.

Name: _____   **Date:** _____

Numbering Squares Arrays

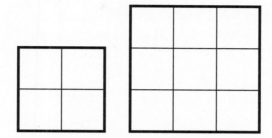

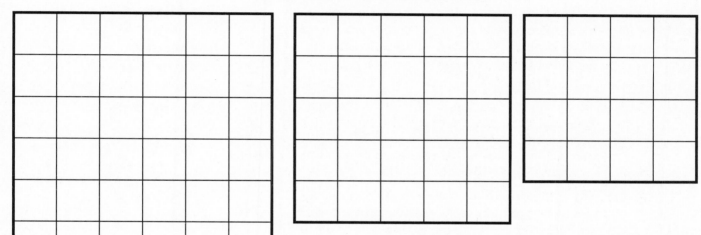

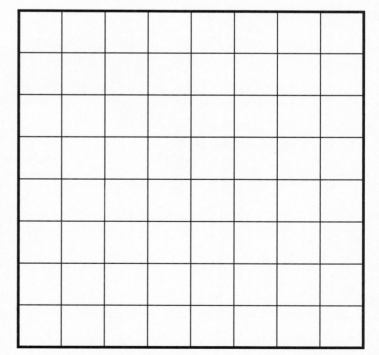

What patterns did you discover? _____

Roll and Graph

Introduce children to the concept of gathering data and making a bar graph. In this activity, children repeatedly roll a dot die and record the number rolled to build a bar graph.

NOTE: Early learners may need some direct instruction on how to build a bar graph. Show them where to start recording data in each column of the graph.

HERE'S HOW

Distribute copies of the Roll and Graph recording sheet and a dot die to children. Children can work individually or in pairs. Decide how much time to spend on building the graph.

1. To begin, children roll the die, then write the number rolled in the matching column on their recording sheets.

2. Children keep rolling the die and recording the number until time runs out.

3. Gather children together to compare and discuss the results of their graphs.

MATERIALS

- **Roll and Graph recording sheet (page 39)**
- **dot die**
- **pencils**

VARIATIONS

- More advanced children can use two dot dice and a bar graph labeled with numbers 1 to 12 (page 40) or use three dot dice and a graph labeled up to 18.

- Remote learners can use a 1–6 spinner (available online; see page 4 for the link) instead of a die.

Name: _____   Date: _____

Roll and Graph
(with one dot die)

Roll the die. Write the number you roll in the matching column on the graph. Start recording at the bottom of the graph. As you roll more numbers, fill in the graph going up the columns.

1	**2**	**3**	**4**	**5**	**6**

Name: _____ Date: _____

Roll and Graph
(with two dot dice)

Roll the dice. Write the number you roll in the matching column on the graph. Start recording at the bottom of the graph. As you roll more numbers, fill in the graph going up the columns.

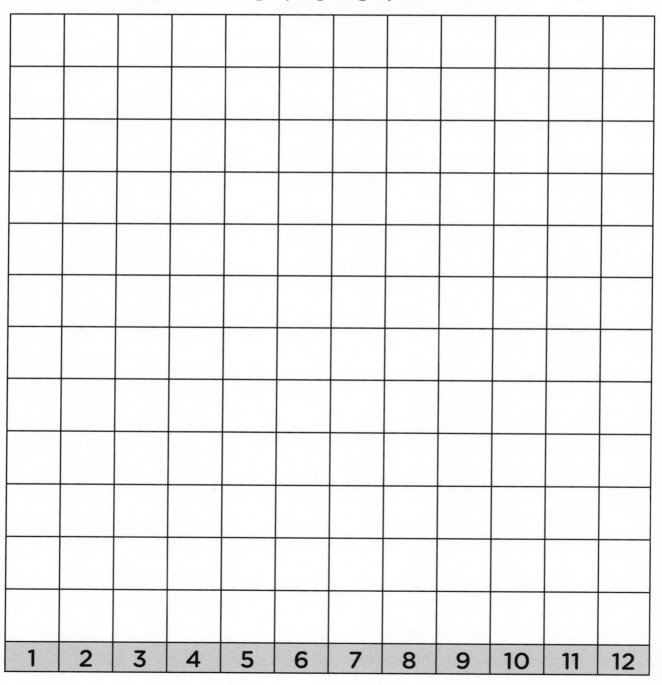

| 1 | 2 | 3 | 4 | 5 | 6 | 7 | 8 | 9 | 10 | 11 | 12 |

Toothpick Patterns

Building simple, identical shapes is a good way for children to visualize growing patterns, including patterns created by skip counting or counting in multiples. Toothpicks are a simple, inexpensive tool to use for building shapes.

HERE'S HOW

1. Decide on a simple shape for children to build (see below for suggestions).

2. At the top of a sheet of paper, have children build the shape using toothpicks.

3. Below that first shape, have them build two of the same shape. Make sure the shapes are distinct and don't share any sides. This is the second row.

4. Have children continue with the activity, building three shapes on the third row, four shapes on the fourth row, and so on.

5. They can glue the toothpicks to the paper.

6. Afterwards, have children count and record the total number of sides in each design.

Try these shapes:

- Triangles (count 3, 6, 9, 12, ...)
- Squares (count 4, 8, 12, 16, ...)
- Pentagons (count 5, 10, 15, 20, ...)

MATERIALS

- paper
- toothpicks
- glue

VARIATIONS

- For more advanced children, allow them to create shapes that share sides and count the toothpicks. Try these shapes:
  - Triangles that share sides (count 3, 5, 7, 9, ...)
  - Squares that share one side (count 4, 7, 10, 13, ...)
  - Pentagons that share one side (count 5, 9, 13, 17, ...)
- 🖥 For remote learners, children can use paper and pencil to draw the shapes and label them to count multiples.

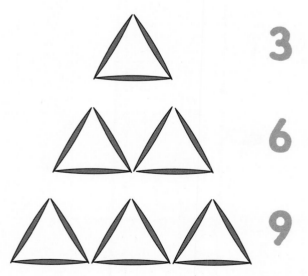

Growing Patterns With Linking Cubes

A growing pattern is one that increases with every new item added to the original shape. The shape changes every time the pattern repeats. We can create growing patterns using shapes, numbers, colors, and so on. When children build growing patterns with linking cubes, be sure to have them identify where and how the shape is changing.

Play & Learn Math: Number Sense © Susan Andrews Kunze. Scholastic Inc.

HERE'S HOW

An effective way to introduce growing patterns to young learners is to provide instruction using an "I do, you do" approach.

1. Walk children through the process of building various growing patterns using linking cubes (see below for some examples).

2. Start by identifying the unit or units within the original shape. Then add a cube to each changing section within the shape, while asking children to predict what the complete shape will look like.

3. On a T-chart, record the number of cubes you added and the total number of cubes in the shape to help children identify interesting number patterns created by the growing shape pattern.

MATERIALS

• linking cubes
• paper and pencil

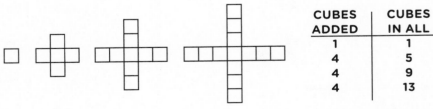

CUBES ADDED	CUBES IN ALL
1	1
1	2
1	3
1	4

CUBES ADDED	CUBES IN ALL
1	1
2	3
3	6
4	10

CUBES ADDED	CUBES IN ALL
1	1
4	5
4	9
4	13

CUBES ADDED	CUBES IN ALL
1	1
3	4
5	9
7	16
9	25

Grab Some Beans

This simple activity provides children with opportunities to understand number quantities and even learn the commutative property of addition. The repetition and math discussions that happen as they play give children practice in an authentic context.

HERE'S HOW

Provide each group of three or four children a container of beans (or any small object) and a dot die.

1. One player rolls the die to determine the number of beans to take.

2. He takes that number of beans and puts them in both hands in any combination. (For example, if the player rolls a 4, he may hold one bean in one hand and three in the other. Or he can put all four beans in one hand.)

3. The player opens one hand to show the other players. All children say aloud the number of beans in that hand.

4. Keeping the first hand open, the player opens the other hand. Players say the number of beans in each hand. (For example, when the player opens one hand, the other players say, "One." When he opens the second hand, the other players say, "One and three.")

5. The next player takes the same number of beans, puts them in both hands using another combination, and shares them with the others.

6. Play continues for as many rounds as the number rolled. For example, if a 4 is rolled, play continues for four rounds, with each player making a new combination for 4 in each round.

7. Then another player rolls the die to determine the next total of beans to be used.

PLAYERS 3 or 4

MATERIALS

(for each group of players)

- **container of beans (or any small objects that can be used for counting)**
- **dot die**

VARIATIONS

- To explore larger numbers, vary the dice used. A dot die allows for developing number understanding for quantities up to six. A decahedron (ten-sided) die allows children to work with numbers up to ten.

- For remote learners, have children go on a scavenger hunt at home to find beans or other small objects for counting. The teacher rolls the die, and children make a combination of beans in their hands to show the number. Discuss and record all the different combinations that add up to the rolled number.

Bean Toss

Using two-color beans is a simple way to generate two addends that combine to a specific sum. Children enjoy shaking cups of beans, keeping them engaged as they work to identify various combinations of number pairs that add up to a single sum. You will need to prepare the dried beans by spray-painting them (see Preparation below).

HERE'S HOW

Partner up children and give each pair a handful of beans, a dot die, a small cup, and paper and pencil.

1. One child rolls the die, counts out that many beans, and places them in the cup.

2. Placing a hand over the mouth of the cup, the child shakes the beans then tosses them on a flat surface.

3. Partners work together to count the number of beans with the painted-side up and the number of beans with the unpainted-side up.

4. The other child writes an addition sentence that matches their bean amounts.

5. Partners switch roles and repeat the process, recording each addition sentence as they play.

VARIATION

To work with sums greater than 6, give each pair a decahedron (ten-sided) die or two six-sided dot dice. Ready for larger numbers? Use three dot dice to sum to 18.

MATERIALS

(for each pair of children)

- 2-colored beans (see Preparation)
- dot die
- small cup
- paper and pencil

PREPARATION

You will need a package of dry beans (lima or kidney beans work well) and a can of spray paint. Out in open air, lay the beans flat on sheets of newspaper and spray paint only one side. Let them air dry. (Alternatively, you can purchase commercially available two-color "bean" counters.)

Subtraction Trains

Children use linking cubes to build equal-length trains, then roll a die to determine how many cubes to remove from their trains. Using this manipulative tool is a great way to demonstrate the "take away" form of subtraction.

HERE'S HOW

Partner up children or put them in small groups of up to four children. Give each group several linking cubes, a dot die, and pencil and paper.

1. Each player uses the linking cubes to make a train. All players' trains must be equal in length.

2. Players take turns rolling the die and removing that many cubes from his or her train.

3. They then write the subtraction sentence that matches their move. (For example, if a train is ten cubes long and a player rolls a 3, she breaks off three cubes from her train and writes "10 – 3 = 7.")

4. Play continues until one player removes all cubes from his or her train. The first player to remove all cubes wins.

5. Players can start the game again, building new linking-cube trains.

PLAYERS   2 to 4

MATERIALS

(for each group of players)

- **linking cubes**
- **dot die**
- **paper and pencil**

VARIATIONS

- Players may decide that only the exact number rolled may be removed from a train. So, if a player has two cubes left, she will have to roll a 2 to remove the cubes. This will extend the time needed to finish a game as it nears completion.

- The teacher designates the length of the linking-cube trains.

- For enrichment, have children compare the length of their trains after each turn, then describe or write that difference as a comparison subtraction sentence. For example, if one player has seven cubes left in his train and another has nine left in her train, they could say that the second player has two more cubes than the first player or write 9 – 7 = 2.

Many Ways to Say It

Children become increasingly nimble with numbers through activities that challenge them to identify equations with the same sum or difference. Using dominoes and playing cards, this game encourages children to find as many addition and subtraction sentences to express the same number.

HERE'S HOW

Place the dominoes faceup between the players. Shuffle the Number Cards and stack them facedown.

1. Each player takes three Number Cards and lays them faceup in a row. If a player has two or more cards of the same number, he returns the extra card(s) to the bottom of the stack and picks a new card.

2. Players find as many dominoes as they can with a sum or difference that matches the number on each of their cards. (For example, a domino with five and two pips matches either 7 or 3.) They place the appropriate dominoes in a column below each Number Card.

3. When players can find no more matching combinations, they switch places and check each other's dominoes for accuracy. (You may want to provide counters for children to use, in case of disputes.)

4. Afterwards, players count the number of dominoes they used for all three of their cards. The player with the most dominoes scores a point.

5. Players clear their space and start another round by picking three new cards.

6. Play continues until time runs out. The player with the most points at the end of the game wins.

VARIATIONS

• To challenge more advanced players, use double-nine dominoes. Use index cards or cardstock to make the numbers 13 to 18.

• Support struggling learners by having two partners play as a team.

💻 For remote learners, introduce the game by demonstrating how to play it. Children can then play alone or with a partner, such as a caregiver or an older sibling, using homemade number cards or playing cards with the jacks and queens representing 11 and 12, respectively.

PLAYERS 2 or 3

MATERIALS

• **Number Cards*** (page 64)
• **double-6 dominoes**
• **counters or other counting objects (optional)**

* For durability, photocopy the Number Cards on cardstock. Laminate and cut the cards apart. You will need one set of cards for each player.

Play & Learn Math: Number Sense © Susan Andrews Kunze, Scholastic Inc.

Say It Again

In this activity, children play with manipulatives to develop a conceptual understanding of numbers and the various ways they can represent a number. Using linking cubes in two colors, children work in pairs to build stacks of a given length in different color arrangements. They then describe them using number sentences.

HERE'S HOW

Partner up children and give each pair with a dot die, linking cubes in two colors, and a whiteboard and dry-erase marker (or paper and pencil).

1. One child rolls the die to determine the target number.

2. Partners work together to build that number using one or two colors of linking cubes. They don't have to keep all cubes of the same color together in a stack. Encourage them to vary the order of the colors. For example, a stack representing 6 might show two yellow, two blue, and two yellow.

3. Children continue building the number in as many different stacks of linking cubes as possible, without repeating the color pattern.

4. On the whiteboard or a sheet of paper, children write a number sentence to describe each stack. For instance, the above example would be 2 + 2 + 2; another stack could be one blue, two yellow, two blue, and one yellow (1 + 2 + 2 + 1); and yet another stack could be six yellow and zero blue (6 + 0).

5. If time allows, have children roll the die again to pick a new number to build.

MATERIALS

(for each pair of children)

- **dot die or decahedron (10-sided) die**
- **linking cubes in 2 colors**
- **whiteboard and dry-erase marker (or paper and pencil)**

NOTE: Children may not recognize two stacks with the same pattern when oriented differently, for example, upside down. If children show confidence in identifying number sentences, you may consider pointing this out and discuss the commutative property of addition.

VARIATION

Remote learners can use toy building bricks or blocks instead of linking cubes. The teacher can roll the die to get a number and children can build the number with their building bricks.

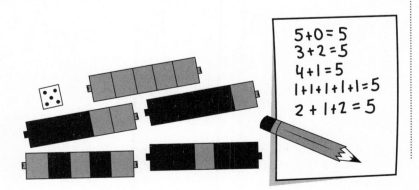

$$5 + 0 = 5$$
$$3 + 2 = 5$$
$$4 + 1 = 5$$
$$1 + 1 + 1 + 1 + 1 = 5$$
$$2 + 1 + 2 = 5$$

Domino Flip

How can you make repetitive math practice feel new? Use a variety of tools, such as dominoes! Played in pairs, this game provides children with engaging practice in early addition and subtraction.

HERE'S HOW

Partner up children and provide each pair with a set of double-six dominoes (or double-nine, depending on their skill level) and a penny. Give each player paper and pencil.

1. Place the dominoes facedown between players.

2. One player flips the penny. "Heads" means write an addition sentence, and "tails" means write a subtraction sentence.

3. Each player picks a domino and writes a number sentence using the two numbers on their domino.

4. Players then trade papers and check each other's number sentence. If the number sentence is correct, the player gets a point.

5. Tally points for each player.

6. Play continues until one player earns ten points. Extra time? Start a new game!

MATERIALS

(for each pair of players)

- double-6 or double-9 dominoes
- penny
- paper and pencil

VARIATION

Make play more challenging by adding a second penny toss. In this toss, "heads" means the greater sum or difference gets the point, and "tails" means the lesser sum or difference gets the point. Adding this randomization of gaining points makes the task more interesting to children.

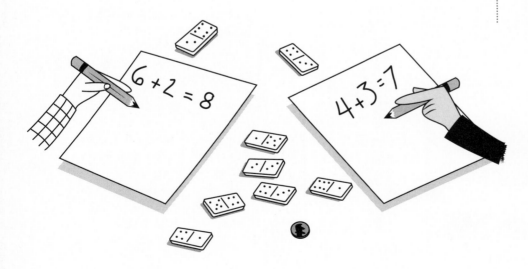

Play & Learn Math: Number Sense © Susan Andrews Kunze, Scholastic Inc.

Cards Make 10

In this simple card game, players race to make sums of 10 from available number cards.

While we provide ready-to-reproduce Number Cards at the back of this book, many teachers already have sets of store-bought playing cards on hand. Since most card games used for learning require only the digit cards, consider breaking up decks of playing cards and put only the ace (1) to 9 cards into a container within children's easy reach. (Save the other cards for those times you might need cards to represent 10, 11, or 12.) This way, children can quickly grab a handful of cards for a small stack to start games quickly!

HERE'S HOW

Partner up children and give each pair four sets of Number Cards.

1. Shuffle the cards and stack them facedown between players. Lay four cards faceup in a row between players.

2. Each player tries to make a sum of 10 using as many of these cards as needed.

   • As soon as a player finds a way to make 10, she takes those cards from the row and keeps them. She then refills the row from the stack of cards.

   • If a sum of 10 cannot be made from the four cards, players can remove one card and return it to the bottom of the stack. They then take a new card from the top of the stack and put it in its place so play can resume.

3. Play continues with players taking sums of 10 as they discover them.

4. The player with the most cards at the end of the game wins.

PLAYERS 2

MATERIALS

• **Number Cards, 1–9*** **(page 64)**

* Photocopy four sets of Number Cards on cardstock for each pair of players. Laminate and cut the cards apart.

VARIATION

For remote learners, introduce this game by demonstrating how to play it. Children can then play alone or with a partner, such as a caregiver or an older sibling, using homemade number cards or playing cards with 10 and the face cards removed.

Toss and Add

This variation of "Toss and See" (page 24) provides active practice in simple addition strategies. Grab some hula hoops and beanbags from your phys-ed materials, and watch children's addition skills soar.

HERE'S HOW

Lay the hula hoop on the floor and use masking tape to mark a throw line several feet from the hoop.

1. Player 1 rolls the die to determine how many beanbags to use.

2. Player 1 stands behind the throw line and gently tosses all the beanbags toward the hoop.

3. Player 1 counts the beanbags inside and outside the hula hoop and tells the other players a number sentence that describes the results. (For example, if he rolls a 7 and lands five beanbags inside the hoop and two outside, he may say, "5 + 2 = 7.") The others check for accuracy.

4. On a sheet of paper, Player 1 tallies the number of beanbags that landed inside the hula hoop.

5. The next player takes a turn, repeating steps 1 to 4.

6. The game continues with players taking turns rolling the die, tossing the beanbags, saying the matching number sentence, and tallying the number of beanbags inside the hoop. The first player to score 25 points wins.

PLAYERS 3 or 4

MATERIALS

(for each group of players)

- hula hoop*
- masking tape
- decahedron (10-sided) die
- 10 beanbags
- paper and pencil

* No hula hoops? Use a long piece of yarn to form a circle on the floor. If you don't have beanbags, try soft whiteboard erasers, large pink rubber erasers, or clean socks rolled into balls.

Shut the Box

The traditional "Shut the Box" game is played with numbered tiles attached to a wooden box. After tossing two dice, a player "shuts" or covers the number tiles that add up to the sum shown on the dice. The goal is to shut down all the number tiles.

This version of the game uses number cards instead of numbered tiles. Encourage players to strategize which operation—addition or subtraction—to use to cover or turn over the cards.

HERE'S HOW

1. Place the Number Cards faceup in a row, ordered from 1 to 12.

2. Player 1 rolls both dice. Player 1 can choose to add or subtract the numbers on the dice. She then turns over the number card that matches the sum or difference.

3. Player 2 takes a turn. If there is no faceup card with the matching sum or difference, then the player turns over that facedown number card to make it face up again.

4. Players continue taking turns rolling the dice, adding or subtracting the numbers, and turning over cards until they are all facedown.

PLAYERS 2

MATERIALS

(for each pair of players)

- **Number Cards (page 64)**
- **2 dot dice**

VARIATION

For remote learners, introduce this game by demonstrating how to play it. Children can then play alone or with a partner, such as a caregiver or an older sibling, using homemade number cards or playing cards with jacks and queens representing 11 and 12, respectively.

Skip-Counting Trains

An important concept for developing number sense is recognizing that several objects can be represented as one group and one group can be made up of several objects. In this activity, children work in pairs to create equal groups of objects (linking cubes), then skip count to find the total number of objects. This activity also provides manipulative practice for children in counting by numbers other than 1.

HERE'S HOW

Partner up children and give each pair two dot dice and several linking cubes.

1. One child rolls a die to see how many cubes to use to make a "train."

2. The other child rolls the second die to see how many trains to make. (For example, if the first child rolls a 3, then they will use three linking cubes to build each train. If the second child rolls a 4, that means they should make four trains in all.)

3. Partners work together to make that many trains by linking together the specified number of cubes.

4. When they finish making their trains, partners skip-count by the number of cubes used to make each train to find out how many cubes there are in all. (In the example above, children would count, "3, 6, 9, 12.")

5. Have children repeat steps 1 to 4 to make more trains.

MATERIALS

(for each pair of children)

- **2 dot dice**
- **linking cubes**

VARIATIONS

- If children are ready to work with larger numbers, have them use decahedron (ten-sided) dice to increase the challenge.

- 🖥 Remote learners can use toy building bricks or blocks instead of linking cubes. The teacher can roll one die to determine the number of cubes or blocks to use and let each child roll his or her own die to determine how many trains to build. No dice at home? Use a 1–6 spinner instead (available online; see page 4 for the link).

Breaking Trains

Breaking apart a "whole" into smaller, equal groups helps children develop a conceptual understanding of division. With this activity, children subtract a number repeatedly from the whole, and there may or may not be a remainder. Encourage math talk as children discover by trial and error which numbers break evenly and which have remainders.

HERE'S HOW

Have children work together in small groups of three or four. Give each group several linking cubes and a dot die.

1. Have each child in the group make a long "train" of linking cubes. All the trains should be the same length.

2. One child rolls the die, and everyone removes groups of the rolled number from their trains. (For example, if the number rolled is 3, children remove groups of three cubes from their trains.) They stack the groups in front of them.

3. When finished, each child counts how many groups she made and how many cubes are left over from her train. Since all trains are the same length, each child should have the same number of groups and leftovers.

4. Have children reassemble their trains and repeat steps 2 and 3 five more times, with children taking turns rolling the die. If a child rolls a number that has appeared before, have her roll again.

5. Have children continue rebuilding trains of the same length and breaking them apart until they have rolled five different numbers.

6. Afterwards, have children make new trains of a different length, making sure all their trains are equal. Repeat the activity.

MATERIALS

- linking cubes
- dot die
- Breaking Trains recording sheet (page 54), optional

VARIATIONS

- Have children use the Breaking Trains recording sheet to keep track of equal groups and remainders. For some children, recording data can guide them to make new mathematical discoveries.

- Remote learners can use toy building bricks or blocks instead of linking cubes. No dice at home? Use a 1–6 spinner instead (available online; see page 4 for the link).

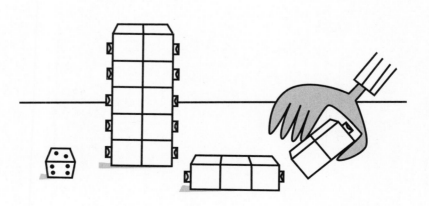

Name: _____ Date: _____

Breaking Trains

Length of train	Break into groups of . . .	Does it break evenly?	How many are left over?

Play & I earn Math: Number Sense © Susan Andrews Kunze, Scholastic Inc.

Count and Group Tens

One of the basic concepts young children need to master in order to understand place value is that ten objects can be put together to form one group of ten and that one group of ten can be separated into ten single objects. Counting and putting together groups of ten objects is an easy and effective way for children to develop an understanding that a group made up of several objects can be considered as a "one."

HERE'S HOW

1. Provide children with various small objects to practice counting and making groups of ten. Many easy-to-find objects work well, including paper clips, beans, toothpicks, coffee stirring sticks, buttons, and pennies. Small containers, such as condiment or soufflé cups, can hold groups of ten objects.

2. Encourage children to count larger objects as well. Counting the books on a classroom bookshelf and using an index card to mark groups of ten books makes use of a ready-made collection to practice grouping. Children can also group playground balls, beanbags, and jump ropes. Take a look around your classroom to find other things children can count and group. Provide many experiences with different types of objects to help them gain foundational understanding of place value.

3. Another way to practice counting and grouping is to have children draw a simple shape within a given amount of time, then circle groups of ten when time is up. Photocopy and distribute any of the following reproducible pages—"Count the Stars!" or "My First Letter"—to offer children some quick draw-and-count activities. There's also a fill-in-the-blank reproducible for you to insert any shape you want children to draw and count.

MATERIALS

- small objects, such as paper clips, beans, or buttons
- small containers for holding objects
- Count the Stars! (page 56)
- My First Letter (page 57)
- Count It! (page 58)

VARIATION

For remote learners, have children go on a scavenger hunt to find small objects (such as those listed above) in their homes. They can use these to count during your online lesson.

Name:_____ Date:_____

Count the Stars!

Draw as many stars as you can in one minute. Then circle groups of ten stars. Record the number of groups and leftover stars below.

Groups of ten stars: _____   Leftover stars: _____

My First Letter

What is the first letter of your name? Write it as many times as you can in one minute. Then circle groups of ten letters. Record the number of groups and leftover letters below.

Groups of ten letters: _____ Leftover letters: _____

Name: _____ Date: _____

Count It!

Draw as many _____ **as you can in one minute. Then circle groups of ten. Record the number of groups and leftovers below.**

Groups of ten: _____ Leftovers: _____

Count and Compare

In this game, children connect a manipulative representation of a number with its numerical symbol. This is an important step in developing understanding of any mathematical concept, particularly recognizing, building, and comparing groups of tens and ones.

HERE'S HOW

Partner up children and give each pair linking cubes, two dice, a penny, and paper and pencil.

1. Players make several "trains" of ten linking cubes each to represent tens. Other cubes will be used as ones. Place all cubes to the side to be used as the game is played.

2. Player 1 rolls the dice and places them at the top of the workspace. One number die represents the number of tens, and the other represents the number of ones.

3. Using the trains and cubes, Player 1 builds that number on the workspace below the dice and names the number he built.

4. Player 2 takes a turn, repeating steps 2 and 3.

5. One player tosses the penny. "Heads" means the player with the greater number wins a point. "Tails" means the player with the lesser number earns a point. If both players built the same number, both players get a point.

6. Players tally their scores on a sheet of paper.

7. The game continues with players taking turns rolling the dice, building and naming their numbers, and comparing them. When time is up, the player with the most points at the end of the game wins.

PLAYERS 2

MATERIALS

(for each pair of players)

- 2 dot dice
- linking cubes
- penny
- paper and pencil

VARIATION

To provide more support for early or struggling learners, give each player two dice to use. After rolling, the player leaves the dice above the trains and cubes to show the actual number with the representation.

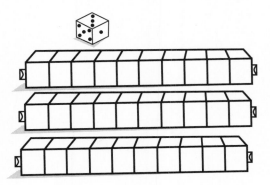
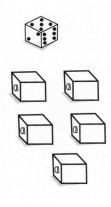

Toss-Up Sticks

Toss-Up Sticks offer a visual representation of a group of ten as one item. These useful tools help children develop early place-value understanding. It takes some preparation, but once you make a class set of Toss-Up Sticks, you can use them year after year for learning fun!

HERE'S HOW

Make a set of nine popsicle sticks for each child. (Consider making extra sets, if possible.) Each stick will have a group of ten dots on one side and one dot on the flip side.

1. On one side of each stick, draw ten dots in a row. Use a color marker to draw the first five dots and a different-color marker to draw five more dots. (Using their subitizing skills, children can quickly identify that the two groups of five dots make up a group of ten.) Try to center the ten-dot design on the length of the stick.

2. On the flip side of each stick, use the third color marker to draw one dot in the center.

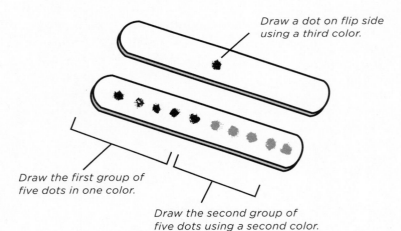

Draw a dot on flip side using a third color.

Draw the first group of five dots in one color.

Draw the second group of five dots using a second color.

3. Store each set of nine sticks in a plastic storage bag or wrap them together with a rubber band. Now they are ready to use!

Children will use the Toss-Up Sticks in the games "Tens and Ones Toss-Up" (page 61) and "Make That Number!" (page 62). These games provide engaging classroom practice in identifying the values of the tens and ones digits in two-digit numbers.

MATERIALS

- wooden popsicle sticks (9 per child)
- 3 medium-point permanent markers of different colors
- rubber bands or plastic storage bags to hold each set

Tens and Ones Toss-Up

This is one of my students' favorite games. Using the Toss-Up Sticks (page 60), children get fun and engaging practice in early place-value skills by identifying two-digit numbers and comparing their values.

HERE'S HOW

Partner up children and give each player a set of Toss-Up sticks. Give both players a penny and paper and pencil.

1. Players gently toss their sticks and name the number they tossed. For example, if four sticks land with the ten-dot side faceup and five sticks land with the one-dot side faceup, the player says, "Four tens and five ones make 45." Or if three sticks land with the ten-dot side faceup and six sticks land with the one-dot side faceup, the player says, "Three tens and six ones make 36."

2. Players identify who has the greater number and the lesser number.

3. One player tosses the penny. "Heads" means the player with the greater number wins a point. "Tails" means the player with the lesser number earns a point. If both players got the same number, they both get a point.

4. Players tally their scores on a sheet of paper.

5. Play continues until time is up. The player with the most points at the end of the game wins.

PLAYERS   2

MATERIALS

• set of Toss-Up Sticks for each player (page 60)
• penny
• paper and pencil

VARIATION

For remote learners, send home a set or two of the Toss-Up Sticks and introduce this game by demonstrating how to play it. Children can then play with a partner, such as a caregiver or older sibling.

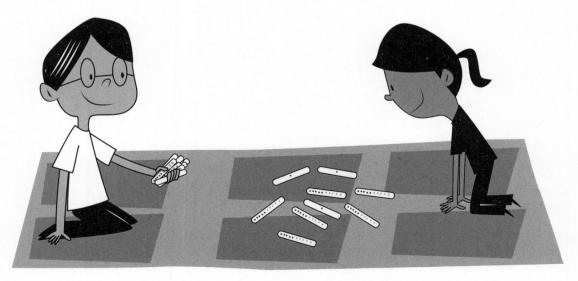

Make That Number!

Children use the Toss-Up Sticks to make representations of two-digit numbers, which they record and compare the numbers of tens and ones. This is another engaging activity designed to help children develop place-value understanding.

Partner up children and give each player a set of Toss-Up Sticks and paper and pencil. Give each pair a small stack of Number Cards. Shuffle the cards and stack them facedown.

1. Player 1 takes two cards from the stack and lays them faceup on the table to make a two-digit number.

2. Player 1 then builds that number with his Toss-Up Sticks.

3. On a sheet of paper, Player 1 writes her number. Next to the number, she draws a representation of tens and ones, similar to this:

4. Player 2 takes a turn, repeating steps 1 to 3.

5. Players continue taking turns picking cards, building the numbers, and recording their amounts as long as time allows.

VARIATIONS

- To provide additional challenge, give each pair of players a penny to toss after each round to keep a point tally. "Heads" means the player with the greater number of tens gets a point, and "tails" means the player with the greater number of ones gets a point. Tie? Both players get a point. The player with the most points at the end of the game wins.

PLAYERS 2

MATERIALS

(for each player)

- set of Toss-Up Sticks (page 60)
- Number Cards, 1–9* (page 64)
- paper and pencil
- penny (optional)

* Photocopy a class set of the Number Cards 1–9 (one set for each child) onto cardstock. Laminate and cut the cards apart.

🖥 For remote learners, send home a set of the Toss-Up Sticks and introduce this game by demonstrating how to play it. Children can then play alone or with a partner, such as a caregiver or older sibling, using homemade number cards or playing cards with 10 and the face cards removed.

Dot Cards 1–12

Number Cards 1–12

1	2	3
4	5	<u>6</u>
7	8	<u>9</u>
10	11	12